Francis Bacon And The Modern Dilemma

92 B1286e   63-21047
E.seley
Francis Bacon and the modern
dilemma

92 B1286e   63-21047
E.seley $3.00
Francis Bacon and the modern
dilemma

FRANCIS BACON
AND THE MODERN DILEMMA

Francis Bacon
and the Modern Dilemma

Loren Eiseley

UNIVERSITY OF NEBRASKA PRESS

Lincoln   1962

To
Francis Bacon
and Sir Thomas Meautys
his faithful secretary
who erected his monument
and chose in death
to lie at his feet
sharing honor and disgrace
this tribute
from one who
more than three centuries beyond their grave
is still seeking
the lost continent of their dream

CONTENTS

THE MAN
WHO SAW THROUGH TIME

1.

The Man Who Saw
Through Time

In 1620 the Pilgrims landed upon Plymouth Rock. Fifteen years before, just prior to the time the Jamestown colonizers boarded ship, there was published in England a book whose author had been powerfully influenced by those western waters over which a few voyagers were beginning to find their way by primitive compass into a new and hidden world. An early edition of that book—*The Advancement of Learning*—lies before me as I write. Sir Francis Bacon, its author, was an Elizabethan stay-at-home destined all his life to hear the rumble of breakers upon unknown coasts.

The book from my shelves is an early edition published in 1640, just fourteen years after Bacon's death. It is crinkled with age and touched by water, and its pages are marked by the rose of a creeping fungus. It has passed through the English civil wars, and Cromwell's cavalry has ridden hard in the night by its resting place. It has been read and dismissed and pondered again by candle, by woodfire, by gaslight. Somewhere in the more than three hundred years of its journey it has crossed the western ocean that moved so powerfully the mind of its creator. There are pages so blackened that one thinks inevitably of the slow way that the fires in the brain of genius run on through the centuries, perhaps to culminate in some tremendous illumination or equally unseen catastrophe.

Not all men, like Sir Francis Bacon, are fated to discover an unknown continent, and to find it not in the oceans of this world but in the vaster seas of time. Few men would seek through thirty years of rebuff and cold indifference a compass to lead men toward a green isle invisible to all other eyes. "How much more," he wrote in wisdom, "are letters to be magnified, which as ships pass through the vast seas of time, and make ages so distant to participate of the wisdom, illuminations, and inventions, the one of the other. . . ." "Whosoever shall entertain high and vaporous imaginations," he warned, "instead of a laborious and sober inquiry of truth, shall beget hopes and beliefs of strange and impossible shapes." It is ironic that Bacon, a sober propounder of the experimental method in science—Bacon, who sought so eloquently to

4

give man control of his own destiny—should have contributed, nevertheless, to that world of "impossible shapes" which surrounds us today.

Appropriately there lingers about this solitary time-voyager a shimmering mirage of fable, an atmosphere of mystery, which frequently closes over and obscures the great geniuses of lost or poorly documented centuries. Bacon, who opened for us the doorway of the modern world, is an incomparable inspiration for such myth-making proclivities. Rumor persists that he did not die in the year 1626 but escaped to Holland, that he was the real author of Shakespeare's plays, that he was the unacknowledged son of Queen Elizabeth. Rumor can go no further; it is a measure of this great discoverer's power to captivate the curiosity of men—a power that has grown century by century since his birth in 1561. In spite of certain mystifying aspects of his life, there is no satisfactory evidence sufficient to justify these speculations, though a vast literature betokens their fascination and appeal.

Perhaps we do not really wish to acknowledge that time and misfortune can fall upon men who, to our more humble minds, seem veritable gods. Perhaps mankind has staked in such geniuses its spiritual effort to evade extinction. The year of 1961, with all its threatened and actual violence, a year of rockets, thinking-machines, and words passing us invisibly upon the air, marked the four hundredth anniversary of the seer who first formulated the vision of our time and who, perhaps more than any other man, set us consciously upon the road of modern science. "This is the foundation of all:" he wrote simply, "for we are not to imagine or suppose, but to discover, what nature does or may be made to do." Bacon defined for us, in addition, the image of what a true scientist should be: a man of both compassion and understanding. He whispered into our careless ears that knowledge without charity could bite with the deadliness of a serpent's venom.

As more and potentially deadlier satellites wing silently overhead, we know at last that Francis Bacon's words were true, though his warning has passed unheeded through three short centuries of unremitting toil which have altered the face of the planet, consumed the green forests of America, and now, at last, have forced the gateway of space which the average Elizabethan regarded as the Empyrean realm of Deity never to be penetrated by man.

"I leave my name," wrote Francis Bacon, lord chancellor of England under James the First, "to the next ages, and the charity of foreigners, and to mine own countrymen after some little time be passed."

6

These words were written in the knowledge that his voice had passed largely unheard in his generation and that, in addition, a calculated fall from royal favor which he had been made to suffer had cost him his good name. In the Tudor century of Elizabeth and the Stuart reign which followed it, the absolute right of monarchs brought many an innocent man and woman to the headsman's block or to forced, communist-like confessions of guilt, as state expediency demanded.

With this kind of atmosphere Francis Bacon was all too familiar. Condemned to spend his life in a world of sometimes petty, often dangerous court intrigue, he chose to have his actual being in a far different one of his own making. He sought to encourage a world the monarchs he served could not envisage, a world of schools, of research promoted by the state, a world in which more could be done than that which, as he phrased it, could be projected "in the hourglass of one man's life."

It is ironic that this man, who died broken and forlorn in an age he never truly inhabited, was warmly loved by such great literary figures as Ben Jonson but arrogantly dismissed by William Harvey, the medical experimentalist renowned for his discovery of the circulation of the blood, as writing of science "like a

7

Lord Chancellor." To that phrase we shall have cause to return later. Sir Edward Coke, the eminent jurist, inscribed in one of Bacon's books the contemptuous comment:

> It deserves not to be read in Schooles
> But to be freighted in the Ship of Fooles.

James the First, the monarch he served, remarked cynically that Bacon's work was like the peace of God that passes all understanding.

King, jurist, scientific experimenter alike saw nothing where Francis Bacon saw a world, a new-found land greater than that reported by the homing voyagers, the Drakes, the Raleighs of the Elizabethan seas. We of the western continent that stirred his restless imagination should particularly honor him because we are also a part, a hopeful foreshore, of the world he glimpsed in time. His great book *The Advancement of Learning* belongs more truly to us today than it ever belonged to that powerful but imperceptive monarch to whom it was originally addressed. It is our opportunity now to share the ideas and visions of a statesman-philosopher whose influence has transcended, and will continue to transcend, his own century.

Man, Bacon insisted, must examine nature, not the

superstitious cobwebs spun in his own brain. He must ascertain the facts about his universe; he must maintain the great continuity and transmission of learning through universities dedicated—not to the dry husks of ancient learning alone—but to research upon the natural world of the present. Then, and only then, would this second world, this invisible world drawn from man's mind, become a genuine reality, erasing narrow dogmatism and prejudice with a true picture of the universe. Though he sought to combine the discoveries of the practical craftsmen with the insights of the philosopher, Bacon saw more clearly than any of the other Renaissance writers that the development of the experimental method itself, the means by which "all things else might be discovered" was of far more significance than any single act of invention.

It was Bacon's whole purpose, set against the scholastic thinking of medieval times, "to overcome," as he remarks in another of his works, the *Novum Organum*, "not an adversary in argument, but nature in action." Truth, to the medieval schoolmen of the theologically oriented universities, rested upon the belief that reality lay in the world of ideas largely independent of our sense perceptions. In this domain the use of a clever and sophisticated logic for argument, rather than observation of the phenomena of nature was the road to wisdom.

In *The Advancement of Learning,* the *Novum Organum,* and several of his other works, Francis Bacon presented quite another "engine," as he termed it, for the attainment of truth. This engine of inductive logic he opposed to the old way of thinking. On this subject much unnecessary argument has taken place among modern scholars who have contended, first, that there is nothing new about Bacon's use of logic and, second, that as a method to discover truth, it is cumbersome and unworkable. Much of the confusion this controversy has engendered arose because of a lack of understanding of the social forces against which Bacon had arrayed himself and a failure to remember that the enormity of the task Bacon had assumed prevented his completion of it. As a result, the deductive side of his logic was left more or less untouched, although there is clear evidence that he remained aware of it.

Inductive logic—the process of inferring a general law or principle from the observation of particular instances—is, in a sense, diametrically opposed to deduction, in which one reasons from an assumed general principle to the explanation of an individual observation or fact. Both of these principles were known to

10

the ancient world, and in this sense Bacon's logic is not original with him. It is in the use to which he put inductive logic that he strove to break out of the old, unproductive circle of the Aristotelian schoolmen. In essence his argument is as follows: We must refrain from deducing general laws or principles for which we have no real evidence in nature. Instead, because of our human tendency to leap to unwarranted conclusions, we must dismiss much of what we think we know and begin anew patiently to collect facts from nature, never straying far from reality until it is possible through surety of observation to deduce from our observations more general laws.

It can, of course, be argued that without some hint, or idea of what we seek, fact-gathering will take us nowhere. Bacon, however, was contending against a philosophy almost diametrically opposed to the examination of the natural world. He himself was aware of the way the mind runs in ascending and descending order from fact to generalization and back again, "from experiments to the invention of causes, and descending from causes to the invention of new experiments." "Knowledge," Bacon insists specifically in *The Advancement of Learning*, "drawn freshly and in our view out of particulars, knoweth the best way to particulars again. And it hath much greater life for practise when the discourse attendeth upon the example, than when the example attendeth upon the discourse." "We have," he emphasizes, "made too untimely a departure and too remote a recess from particulars"—that is, from the facts of the natural world about us. Indeed, it was just such facts, impossible to be subjectively gathered from the mind alone—fossil bones, superimposed strata, animal and plant distributions—which led, more than two hundred years after Bacon's death, to perhaps the greatest inductive achievement of science—the creation of the evolutionary hypothesis.

Yet Bacon, for all his emphasis on observation, was ahead of his time and writes, indeed, like a modern theoretical physicist when he argues that "many parts of nature can neither be invented—that is, observed—with sufficient subtlety, nor demonstrated with sufficient perspicuity . . . without the aid and intervening of the mathematics."

Interestingly, Bacon was not unaware of the intuitive insights of genius or the fact that time might yield new ways of discovery. Bacon saw, however—and this has been taken as an affront by the sensitive—that reliance upon the sporadic appearance of genius was no sure or sane road into the future. Long before the rise of sociology and anthropology, he had grasped the concept of the cumulative basis of culture and the fact

12

that inventions multiplied in a favorable social environment. Science and its traditions had to be transmitted through the universities, and its efforts had to be publicly supported. By such means humanity could progress more consistently and far better sustain its great intellects than if such intellects were allowed to flounder amidst an unenlightened and hostile populace. Indeed with laboratory research conducted on a high level there would be less distinction between the achievements of able teams of researchers and the occasional attainments of unsupported genius under the handicaps so evident in Bacon's day.

Bacon's observations were perfectly justified. Our entire school system, however faulty it may remain in particulars, is predicated upon Bacon's faith in the transmission of learning and the continuing expansion of research into that dark realm where man, either for human benefit or terror, can increasingly draw the purely latent, the *possible* out of nature, thus supporting Bacon's conception that the investigation of nature should include a division "of nature altered or wrought." What the philosophers of a later century would call the emergence of novelty in the universe is already clear in his mind.

Experiments of light, Bacon insisted, were more important than experiments of fruit. He sought to restore to man something of his fallen dignity, to erase from his mind the false idols of the market place and to regain, by patient labor and research, some remnants of the innocent wisdom of Adam in man's first Paradise. The mind, he contended, had in it imaginative gifts superior to the realities of sixteenth-century life; in fact, to the realities of the world we know today.

Our ethics are diluted by superstition, our lives by self-created anxieties. Our visions have yet to equal some of his nobler glimpses of a future beyond our material world of easy transport, refrigeration, and rocketry. The new-found land Bacon sighted was not something to be won in a generation or by machines alone. It would have to be drawn slowly, by infinite and continuing effort, out of minds whose dreams must rise superior to the existing world and shape that world by understanding of its laws into something more consistent with man's better nature. "Our persons," he observes, "live in the view of heaven, yet our spirits are included in the caves of our own complexions and customs which minister unto us infinite errors and vain opinions if they be not recalled to examination."

Bacon's science, as he formulated it, was to have taken account of man himself and to have studied

14

particularly the ethical heights to which individuals, and perhaps after them the mass, might attain. His "second world" as his late fragment, *The New Atlantis,* indicates, was to be a world of men transformed, not merely men as we know them amidst the machines of the twentieth century. Thus parts of Bacon's dream lie beyond us still.

The man who opens for the first time a doorway into the future and who hears faint and far off, like surf on unknown reefs, the tumult and magnificence of an age beyond his own is confronted not alone with the scorn of his less perceptive fellows, but even with the problem of finding the words to impose his vision upon contemporaries inclined to the belief that the world's time is short and its substance far sunk in decay. To achieve this well-nigh impossible task, Bacon had to take the language of his period and, like the seer he was, give old words new grandeur and significence, blow, in effect, a trumpet against time, darkness, and the failure of all things human.

It was his task to summon the wise men, not for one day's meeting or contention, not to build a philosophy of permanence under whose shadow small men might sit and argue, but to leave, instead, a philosophy forever and deliberately unfinished—not, as he so ably put it, as a belief to be held but rather as a work to be done. His philosophy, for human good or ill, has brought the foreshore of a great and unknown continent before our gaze. On the beaches of this haunted domain we find the footprints of the greatest Elizabethan voyager of all time—a man who sounded the cavernous surges of the darkest sea against which men will ever contend: the sea of time itself.

It is not possible to realize the full magnitude of Bacon's achievement without some knowledge of this age of the scientific twilight—an age when men first fumbled with the instruments of science yet, in the next breath, might consider the influence of stars upon their destinies or hearken to the spells of witchcraft. It was generally held that the human world was far gone in decay. Men who had only lately tasted the fruits of the lost classic learning had their gaze cast backward upon a civilization they felt unworthy to equal. The giant intellects of Greece, the vanished splendor of the Roman empire held the human mind enthralled; the voices of antiquity ruled the present. This spell—for so it seemed to lie upon men's minds—was the result of a belief in a world whose destiny was incredibly short by the scientific standards of today.

Man's fall from the perfect Garden was believed to have infected nature itself. The microcosm, man, had destroyed the macrocosm, nature. The human drama

16

of the Fall and the Redemption was being played out upon the brief stage of a few trifling millennia. Eternity, the timeless eternity of the spiritual world, was near at hand. By contrast the long story of geological change and evolution was unknown. The smell of an autumnal decay pervaded the entire Elizabethan world. Over all that age which now glimmers before us with jewels of thought more wonderful than the gems from some deep-buried pirates' hoard, there was a subdued feeling in men's hearts that the sands in the hourglass were well-nigh run. It was autumn, late autumn, and God was weary of the play.

Men had come a long way down from the ages of the patriarchs recorded in the Bible, or from those philosophies which had been spun under the younger sunlight of Greece. Man was truly caught up in a spell of his own making; the contentions of past philosophers were on his lips; their systems occupied his brain. Bacon, by contrast, saw the wisdom of the ancients "but as the dawning or break of day." "I cannot," he emphasized in *The Advancement,* "but be raised to this persuasion, that this third period of time will far surpass that of the Grecian and Roman learning: only if men will know their own strength and their own weakness both; and take one from the other light of invention, and not fire of contradiction."

17

"Employ wit and magnificence to things of worth," he chided, "not to things vulgar."

Elizabethan craftsmen or lonely seafarers sometimes made discoveries, he noted, but the experiments of calloused hands were often scorned by gentlemen. After all, what did newcomers matter in a world of falling leaves, with even the moon grown old and spots upon its face? In a sense these people lived as only we today live under the shadow of atomic disaster, with a boreboding that the human future is running out. We know a universe of greater antiquity, but we, in a different way than the Elizabethans, know our own sinfulness with a dreadful renewed certainty. It is with us as they thought it was with them. Our civilization smells of autumn. "Lest darkness come," St. John records the remarks of Jesus in the New Testament, "believe in the light, that ye may be the children of light."

Francis Bacon never wavered in the faith that this light was compounded of both knowledge and charity. With the first of these virtues he sought to open up a hidden continent for man's exploitation. Of his second virtue, it is all too evident that it remains today largely disregarded. Yet only by charity and pity did Bacon foresee that man might become fit to rule the kingdom of nature. The technological arts alone had concealed in them, he realized, a demonic element. They could bring men riches, but they could draw out of nature powers which then became non-natural because they were subject to the human will with all its dangerous implications.

Bacon was the unwilling servant of an age he loathed. Yet he has been charged with calling up from the deeps of time that which we, in the modern world, have not found the power to put down. We cannot say Bacon did not warn us. The contemplation of the light of wisdom, he argued, was fairer than all its uses. He pleaded for the instruction of youth in high example. His warning to begin at the very threshold of his new continent the search for protection against its dangers went unregarded.

Bacon knew and said repeatedly that the light of truth could pass without harm over corruption and pierce unsullied the darkest and most noisome sewers. Perhaps in this sense he foresaw how far into the depths of ourselves we must descend in order to vanquish that serpent of evil which, projected into biblical mythology is, in fang, venom, and scale, a considerable part of our long evolutionary heritage and, not incredibly, perhaps, our Nemesis. Yet if Bacon's contemporaries felt the blast of some final autumn at their backs, Bacon, the midnight reader and stay-

at-home, had sensed and transferred to his own peculiar dimension a fresher and more vital wind.

It was something the giants of antiquity had not
mastered—the wind of the world-girdling voyagers.
"This proficience in navigation and discoveries," he
wrote, "may plant also an expectation of the proficience and augmentation of all sciences: because it
may seem they are ordained by God to be coevals, that
is, to meet in one age." Here in this supposed dead
season of the intellect existed a triumph unclaimed by
antiquity. Not for nothing did the first edition of his
Great Instauration show a ship passing in full sail
through the pillars of Hercules; not unknowingly did
Bacon speak of himself as a stranger in his century—
an explorer passing as warily as a spy through foreign
lands.

The voyagers had brought home tales of continents
and men undreamed of within the little confines of
Christian Europe. Bacon sensing almost preternat-
urally the meaning of that oceanic wind, had raised a
moistened finger in what, by contrast, seemed the dead
calm of a few sparse human generations. Against that
finger, though faintly, he had sensed the far-off wind
of the future. He had envisioned man's power to
change and determine his own destiny. Scientifically,
he was one of the first to grasp the latent novelty that

20

could be drawn out of nature. He was beginning to discover history and world time—a phenomenon which the historian Friedrich Meinecke was later to call "one of the greatest spiritual revolutions which western thought has experienced." It meant, as Lord Morley observed astutely, "the posing of an entirely new set of questions to mankind."

In that far continent in time, against whose coasts Bacon caught the murmur of troubled surf, "many things are reserved which kings with their treasures cannot buy, nor with their force command, their spials and intelligencers can give no news of them, their seamen and discoverers cannot sail where they grow." This hidden world, he argued, could be brought out of nature only by a great act of the human imagination. "Reason," he proposes, "beholds a farre off even that which is future." There is possible, he insists, a kind of natural divination, a key to the opening of nature's secrets, rather than the idle acceptance of immobility, of pure dogmatism, of "animal time." The sovereignty of man over nature, Bacon contended, lies in knowledge. "For that which in these perceptions appeareth early, in the great Effects cometh long after." "It is true also," he writes, "that it serveth to discover that which is hid, as well as to foretell that which is to come. . . ."

By now it should be evident what lay hidden in the fallacious contempt of William Harvey when he spoke of Bacon as writing of science like a lord chancellor. This is precisely what Bacon did, but what he lacked as a personal experimenter he made up for in his range and vision of what science in its totality meant for man. Bacon was an experienced planner and adviser upon matters of diplomacy and affairs of state. He was, in fact, the first scientific scholar to approach the incipient institution of science from the viewpoint of a practical statesman.

The individual discoverers would come later. This, with sure instinct, he knew. The real problem was to break with the dead hand of the traditional past, to free latent intellectual talent, to arrest and touch with hope the popular mind, to carry word of that which lay beyond the scope of the isolated individual thinker; namely, to dramatize what we have previously referred to as the invention of the experimental method itself—the invention of inventions, the door to man's control of his own future. "Every act of discovery," Bacon enjoins us, "advances the art of discovery."

Bacon's discoveries lie in the intangible realm of thought. "Deeds need time," wrote Friedrich Nietzsche, "even after they are done, to be seen and heard." The mind of Bacon, whose thoughts are still moving toward and beyond us across the dark void of centuries, already shines with the impersonal radiance of a star.

What we make of Bacon's second world in every human generation lies partly in the unfathomable realm of human nature itself. We may dip into his pages and write of him as one who, unwittingly, set us on the road which led at last to stalemate by terror, to a world divided by a fence of mine fields, barbed wire, and concrete blocks. Just as readily we could establish that he urged from the beginning of science that man's own powers must be used with wisdom. The poet Samuel Taylor Coleridge, a great student of Bacon, was actually echoing the Lord Chancellor when he said that man was a secondary creator of himself and of his own happiness or misery.

In *The Advancement of Learning* Bacon makes clear his concern, not only with knowledge, but its application for human benefit and freedom. He knew that man himself, unless well studied and informed, was part of the darker aspect of that unknown country which, as he said, "awaited its birth in time." "Mere power and mere knowledge exalt human nature but do not bless it," he insisted. "We must gather from the whole store of things such as make most for the uses of life."

For the uses of life! I switch off my reading light for a moment and a knob manipulated by my hand brings from the ends of the earth threat and counter-threat. The wars of the Cavaliers are ended. Men speak to each other now like the wrathful God of the Old Testament, threatening to make of their enemies' countries dust and a habitation for owls. The threats are real, the power, torn from nature, lies exposed in human hands. The voices pass, faintly contending, on their way to the vast silences of space.

For the uses of life. I repeat Bacon's phrase in pain, in the darkness of my study. We who are small, and of those unknown generations for whom Bacon labored and to whom he left his name for judgment, are now ourselves to be judged. Charles Williams, one of Bacon's biographers, has remarked perceptively that Bacon's strange and shadowed life, his proffer of powers from which men shied with reluctance, was the heaviest burden of his genius and his rejected love for mankind. "He brought with him" confessed Williams, in the days before the atomic era, "something that might easily become a terror. Men like Bacon are not easily loved or used: something terrific exists in them, however humbly they speak." "Even to deliver and explain what I bring forward," Bacon once remarked in weariness, "is no easy matter, for things in themselves new will yet be apprehended with reference to what is old." In the passage of long centuries the endless innovations of science have not quieted that lust for power which still blocks the doorway to the continent of Bacon's dreams.

⟨Written into *The Advancement of Learning* in words that should be read and reread by every man who calls himself civilized, is Bacon's key to the true continent. It has always been known to the great seers, but now, and all too rapidly, it must become the property of mankind in general, or mankind will perish. "The unlearned man," wrote Bacon carefully, "knows not what it is to descend into himself, or call himself to account . . . whereas with the learned man it fares otherwise, that he doth ever intermix the correction and amendment of his mind with the use and employment thereof."

"For the uses of life," we might well reiterate, for so he intended, and this is why his green continent lies beyond us still in time. Five little words have shut us, with all our knowledge, from its shores—five words uttered by a man in the dawn of science, and by us overlooked or forgotten.⟩

Francis Bacon coughed out his life alone in a cold and borrowed bed, in a century to which he had been a stranger. He had once written to a friend, "I have

lost much time with this age; I would be glad to re-
cover it with posterity." Bacon saw, like the statesman he was charged with being, the full implications
of science and its hope for man. He left his work unfinished and open to improvement, as he knew science
itself to be unfinished in the world of time. He warned
man of the shadows in his own brain which kept him
dancing like one enchanted within the little circle of
narrow prejudice and fanatic ideology.

Though Bacon died childless his intellectual children in a few scant generations were destined to be
legion. He would be called by Izaak Walton a few
years after his death "the great Secretary of all Nature." He planned as a man of great affairs, dreamed
like a poet, and yet sought to rope those dreams to
earth as though, in doing so, he might more easily sail
earth itself into the full wind of that oncoming and
creative age toward which his vision hastened.

We of today, cognizant of the earth's long past and
our own bestial and half-human origins, grow faint-
hearted at the road we have still to travel. Bacon saw
within our souls the possibility of a return to a Garden, a regaining of innocence through wisdom. We, by
contrast, must seek a garden drawn from our own imperfect evolutionary substance. This is why our
imaginings must be so carefully scrutinized lest we call

the wrong world into being by the dark magic of ill-considered thought.

This is the reason also why Bacon placed such emphasis upon education. The teacher—this oft-scorned handler of young minds—is literally engaged in the most tremendous task imaginable. He is encouraging Bacon's power to create a choice of worlds in the minds of the young.

(Perhaps in the end the Great Bringer of all things out of darkness may smile at our human prejudice, for, fallen angel or risen ape, the end the noblest intellects pursue is still the same: the opening of man's mind to his true dignity. Francis Bacon, the solitary discoverer of a continent in time, has placed his footprint where no man may follow by mechanical invention alone. In prayer Bacon himself had urged "that human things may not prejudice such as are divine, neither that from the unlocking of the gates of sense, and the kindling of a greater natural light, anything of incredulity or intellectual night may arise in our minds towards divine mysteries."

(For us of this age the shadow of that night of which he spoke has fallen across half the world. Perhaps, however, he saw deeper, farther, than ourselves, aware as he was of the "impediments and clouds in the mind of man." "I do not think ourselves," he ventured, with-

27

out the twentieth-century's boastful arrogance, "yet learned or wise enough to wish reasonably for man. I wait for harvest time, nor attempt to reap green corn.")

BACON AS
SCIENTIST AND EDUCATOR

2.

Bacon as Scientist and Educator

In the course of three centuries of examination of Bacon's career, many of the ideas propounded about him have been expressed, forgotten, or re-elaborated by later writers. Kuno Fischer, in the mid-nineteenth century, dwelt upon his marvelous grasp of time. Similarly, Richard Whately perceived that Bacon was a statesman and strategist of science rather than an individual discoverer. Lord Acton, the well-known historian, in the same period, recognized the value in Bacon's promotion of a division of labor in science by means of which discovery need not always wait upon the chance appearance of great genius. He saw the idea as equally applicable in the field of historical research and did not project upon it some of the misplaced animus of scientists jealous of their role as geniuses. Most of these ideas of Bacon's have now been realized in actual practice, but have become so commonplace with the years that his farsighted role in their encouragement has been forgotten as well as the atmosphere of opposition in which they were produced.

Bacon was not a system builder save in his grasp of the potential powers concealed in scientific research. In fact he abjured dogmatic and full-formed philosophies, saying, in the *Novum Organum*: "These I call Idols of the Theatre, because in my judgment all the received systems are but so many stage-plays, representing worlds of their own creation after an unreal and scenic fashion. . . . Many more plays of the same kind may yet be composed and in like artificial manner set forth."

Neither, he warned, was science itself free of tradition and credulity. In this last view he was not mistaken, yet he was tolerant to the point of suggesting, elsewhere, the advisability of collecting and comparing all such philosophies for the grains of truth that might lie in them. He had, as I shall indicate, a clear grasp of the importance of the history of ideas, and in this respect once more forecast a field of thought which is only now receiving the attention it deserves.

As a consequence of this aversion to doctrine light and air blow through his windy and perpetually unfinished building. As Bishop Whately remarked, he labored, unlike the system builders, to make himself useless. He saw his efforts not as primarily theoretical but as a necessary prelude to experiment—a work which he, a solitary man whose hourglass of days was short, could only dimly adumbrate.

The way of science, as someone has observed, is the road of the snail that shineth, though it is slow. Yet in these centuries of weighing and reweighing, detraction and praise, no one seems to have encountered Bacon the anthropologist. This is the more strange because foe and friend alike have been quick to admit that he emerged from an atmosphere where the astute judgment of men and the flow of events were paramount. Bacon himself, perhaps out of bitter self-questioning and disappointment, referred to the world he inhabited as one of shadow rather than of light, yet it must be suspected that it was from the observation of men and custom, out of science practiced "like a Lord Chancellor"—to use once more William Harvey's contemptuous phrase—that Bacon drew his laboratory materials and literally produced his second world, or, rather, laid its foundations in the minds of men.

I have said that his building is huge beyond our imaginations, drafty and unfinished. Like all such monuments of genius, it is never truly of the past. Lights flicker mistily in its inner darkness, stones are still moved about by unseen hands. Somewhere within, there is a ghostly sound of hammering, of a work being done. The work is ours, the building is as we are shaping it, nor would Bacon have it otherwise. Since Bacon was a statesman and a pathfinder, no man quite escapes his presence in the haunted building of science, nor the whispers of his approbation or unease. Occasionally the voice grows louder, as now, in our overtoppling part of the structure. To those who listen, the harsh Elizabethan line strikes once more like surf around the shores of his far-off New Atlantis, warning us of man's double nature and perhaps his fate, for Bacon did not hestitate to write: "Force maketh Nature more violent in the Returne."

It is often asserted, frequently by experimental scientists, that Bacon, at best, was a stimulating popularizer of new doctrines but not a true discoverer. These remarks have always seemed to the present writer an appalling underestimation of Bacon's role in the history of human thought. Ironically, he who defended the technicians has frequently been rejected by them. Such men fail to recognize that mere words can sometimes be more penetrating probes into the nature of the universe than any instrument wielded in a laboratory. They have ignored his vision of the very education of which they are the present-day beneficiaries. They have failed to understand, because he wrote before the rise of a professional jargon, the basic contributions in the realm of the social and biological sciences which have passed imperceptibly into our common body of thought. Bacon's "great machine," his system of induction applied to the natural world about us, has obscured the recognition of much that he observed about human culture and the sociological nature of discovery. The instruments of the mind, he observes justly, are as important as the instruments in the hand. He himself has suffered from the effects of his own observation that "great discoveries appear simple once they are made."

Unfortunately, of no human endeavor is this remark more true than in the domain of abstract thought. A shift of wordage, an alteration in the direction of historical emphasis and the most profound ideas may emerge in a new garb with their parentage forgotten and ignored. Oft-times dismissed or misunderstood by the physical scientist, Bacon's contributions in the social sciences were made, in a sense, too early. By the time these subjects had emerged as recognized disciplines, his far-reaching, anticipatory insights were submerged in a welter of new books and newer phrasing. A few examples in the anthropological domain may illustrate the way in which time erodes

the memory of great achievements in thought. In saying this let us remember that the thinker does not
work in a vacuum. Bacon, too, in spite of an active
and original cast of mind, took in and transmuted
much that came to him from Greek and Roman
sources. Unlike his associates, however, he saw the
Ancients in the dawning light of modernity, not as
Colossi whose achievements the living could not equal,
but, rather, reduced to their true stature as men—men
whose accomplishments the hoarded experience of suc-
ceeding generations might enable us to surpass.

"For at that time," Bacon wrote in *Valerius Terminus,* "the world was altogether home-bred, every
nation looked little beyond their own confines or terri-
tories, and the world had no through lights then, as
it hath had since by commerce and navigation, where-
by there could neither be that contribution of wits
one to help another, nor that variety of particulars
for the correcting of customary conceits [ideas]." In
Bacon's time such a view of the ancients was iconoclas-
tic and heretical to an extent difficult to appreciate
in the intellectual climate of today.

Perhaps one of the most obscurely hidden yet profound insights Bacon possessed revolves around his
discovery of the *mundus alter,* the world unknown.
"By the agency of man," contends Bacon, in what was

then a bold and novel interposition of the human into the natural universe, "a new aspect of things, a new universe comes into view." Characterized in the sea-language of the great Elizabethan voyagers, this new universe is, in reality, what the modern anthropologist calls the world of culture, of human "art" with all its permutations and emergent quality.

Bacon is not content to subsist in the natural world as its exists, nor to drift aimlessly in history. The focal point of all his thinking is action, not—as we have observed previously—system building. His new world to be brought out of time by human ingenuity interpenetrates and is interfused with the natural universe, yet remains a thing apart. Prospero's isle, the New Atlantis, can be brought out of nature only by the magic of the human mind directing and utilizing, rather than contesting against, nature. "The empire of man over things depends wholly on the arts and sciences. For we cannot command nature except by obeying her."

As he proceeds in the *Novum Organum* it is interesting to note that long before the rise of anthropology as a science, he had seen the social dynamic involved in cultural change and divorced it from simplistic explanations based on biology. He says: "Let only a man consider what a difference there is between the life of men in the most civilized province of

Europe, and in the wildest and most barbarous districts of New India [the New World]; he will feel it be great enough to justify the saying that 'man is a god to man,' not only in regard of aid and benefit, but also by a comparison of condition. *And this difference comes not from the soil, not from climate, not from race, but from the arts.*"[1]

Three hundred and forty years ago, in other words, Bacon had already curtly dismissed the racist doctrines that have hovered like an inescapable miasma over following centuries. In addition he had clearly recognized the role of culture and cultural conditioning in human affairs. "Custom," he notes, "is the principle magistrate of man's life." Its predominance "is every where visible. Men . . . do just as they have done before; as if they were dead images and engines moved only by the wheels of custom." It is well, he mused in the *Essays*, to stand upon the hill of Truth and, as one might observe from some point of vantage the waverings of an uncertain battle, to see, through the clean air of reason "the errors, and wanderings, and mists, and tempests in the vale below." "So always," he adds solemnly, "that this prospect be with pity."

In his symbolic treatment of the fable of Orpheus

[1] Italics mine. L. E.

we may observe, as balancing the quotation given above, Bacon's statesman-like recognition of the role played by culture in controlling the otherwise uninhibited behavior of man. Here the harp of Orpheus is Bacon's "magistrate."

> In Orpheus's theatre all beasts and birds assembled, and forgetting their several appetites, some of prey, some of game, some of quarrel, stood all sociably together listening to the airs and accords of the harp, the sound whereof no sooner ceased, or was drowned by some louder noise, but every beast returned to his own nature; wherein is aptly described the nature and condition of men; who are full of savage and unreclaimed desires of profit, of lust, of revenge, which as long as they give ear to precepts, to laws, to religion sweetly touched with eloquence, and persuasion of books, of sermons . . . so long is society and peace maintained; but if these instruments be silent, or sedition and tumult make them not audible, all things dissolve in anarchy and confusion.

The cultural tie—custom, in other words—subdues man to its strange music and holds back the expression of his wilder nature.

If we pursue Bacon's influence upon later scholars in anthropology, we come upon a supreme example which remains oddly undocumented. It is well known that Sir James Frazer, the distinguished British student of primitive magic and religion, makes a great point of the fact that all over the world magical practice seems to resolve itself into two principles of thought: that things once in contact with each other continue to act at a distance, or, second, that like things produce like results, that is, that the imitation, with proper ceremonial, of a natural phenomenon by the magician will initiate, for example, a real rainstorm. Concealed behind these two widespread branches of magic, Frazer believed, was a single spurious "scientific" principle which was common to the primitive and untutored mind. That principle he proclaimed as a mistaken Law of Sympathy, the conception "that things act on each other at a distance through a secret sympathy." Magic, as viewed by Frazer, is thus a kind of falsely conceived science based upon a naive projection of human desire upon the exterior universe.

In the *Sylva Sylvarum,* that curious miscellany to which, as we shall see, Bacon devoted his last years, there is a considered discussion of magic in which Bacon refers to the "operations of sympathy, which the writers of natural magic have brought into an art or precept." Bacon then discusses what Frazer would label Contagious and Imitative Magic, both, of course, distinct products of the underlying principle of Sympathetic magic. I do not wish to detract one iota of credit from the stimulating ideas contained in Frazer's grand old classic, the *Golden Bough.* I should like to observe, however, that I believe Frazer's formulation of the "laws" of magic are drawn, in part, from Francis Bacon. To my knowledge, although Frazer makes a single use of magical example from the *Sylva Sylvarum* —so that it is clear he was aware of Bacon's pioneer precedence—he nowhere indicates Bacon's prior discussion of Frazer's principle "that things act on each other at a distance through a secret sympathy." Yet of this same principle Bacon had written, "of things once contiguous or entire there should remain a transmission of virtue from the one to the other."

Naturally Bacon was involved in other matters and did not refine or draw out of this material all that Frazer in a later century would be able to do. One is a little saddened, however, that Frazer did not see fit to acknowledge the stimulus of that elder scholar whose doom it was to plant seeds which sprang from his work at points so distant in time and space that only the most meticulous would care to acknowledge their indebtedness. Bacon was right in his noble and elevating remark that books, like ships, pass through the seas of time and touch other minds in distant ages. It is troubling to our human wish for remembrance that the launcher of such frail vessels never sees upon what shore they land or whether his precious cargo is carried off by erudite smugglers. Broken by the winds and waves of other ages, what the author painstakingly accumulated may be picked up piecemeal as wreckage and distributed in other holds. His one compensation must lie in the hope that the cargo, if not the master, is saved. In this respect Bacon deserves reexamination. A few examples taken almost at random from his work may prove convincing to those who have tended to dismiss his creative powers.

In *The Advancement of Learning* Bacon, in one of his characteristic axiomatic statements has remarked: "He that cannot contract the sight of his mind as well as disperse and dilate it, wanteth a great faculty." It does not take long to demonstrate that Bacon possessed this rare trait to a unique extent. In the Second Book of the *Novum Organum* he dwells upon the types of insensible change which tend to escape the observation rather than the senses. As he again says so aptly, one of the great drawbacks of untrained sense perceptions is that "they draw the lines of things with reference to man, and not with reference to the universe; and this is not to be corrected except by reason and a universal philosophy." He indicates that much of significance in nature takes place by almost imperceptible progression which the heedless fail to remark. In this observation he was laying the theoretical groundwork which underlies the whole domain of biology from evolution and genetics to embryology. In fact, he not alone dilates his mind to take in this important principle of natural history but, true to his own words, he contracts the principle sufficiently to describe minute experiments which were later carried out by others and very possibly in some cases, under his influence.

He says, and this is the matter of concern to us, that the embryology of plants can be studied through the simple process of inspecting day by day the growth of sprouting seeds. Similarly, he comments, "We should do the same with the hatching of eggs, in which case we shall find it easy to watch the process of vivification and organization, and see what parts are produced from the yolk, and what from the white of the egg, and other things." Malpighi, it will be remembered, published his *De formatione pulli in ovo* in 1673. Fabricius' similar study, *De formatione ovi et pulli*, falls within the time of Bacon's last years but does not predate the publication of *The Novum Organum*. In the *Sylva Sylvarum* one notes another aspect of his "contracted" mind at work. Probing speculatively into the nature of life he notes that "the

Nature of Things is commonly better perceived in Small, than in Great." The investigation of lower and more simple animals, he believes, is more apt to disclose the secrets of life than similar studies made in "Disclosing many Things in the Nature of Perfect Creatures, which in them lye more hidden." Again the modern biologist would be forced to agree.

Jean Rostand, in a recent book, pays tribute to Bacon's pioneer interest in teratology, "of all prodigies and monstrous births of Nature, of everything in short that is new, rare and unusual." Here, argues Bacon, nature in her errors reveals herself unbidden. The knowledge of life's mutations and deflections should aid us in comprehending the mysteries of life. The solution of those mysteries was many generations away, but, once more, that quick impatient brain had probed an instant into the domain of evolutionary change. It is a subject which we can pursue further.

Archibald MacLeish, in a fine poem, "Epistle To Be Left in the Earth," narrates the thoughts of a last, desperate survivor on a dying earth. "I pray you," the unknown survivor writes, "(if any open this writing),

Make in your mouths the words that were our names."

Laboring under the heavy burden of his own mortality, he cries anxiously,

I will tell you all we have learned
I will tell you everything.

The poem then runs off into the disjointed efforts of
a man scrawling, haphazard, a last few facts about this
planet:

It is colder now
there are many stars
we are drifting
North by the Great Bear
the leaves are falling
the water is stone in the scooped rocks.

There is a poignant similarity between this verse
and the real-life creation of Bacon's *Sylva Sylvarum*.
The book was written in haste in his declining years
after all his hopes for new universities, co-workers in
science, and aid from enlightened rulers had been dis-
appointed. But the man, like the individual in Mac-
Leish's poem, still struggled to collect the facts out of
which the new continent should be built. He is aging,
hope is gone, the task looms gigantic. He has no ade-
quate conception of the size of the universe he has
attempted to engage. No one will come to his aid.
The book becomes an almost incoherent babble of
facts drawn both from personal observations and

diverse sources. Moss grows on the north side of trees. Strange fungi spring up in the forests. Fruit put into bottles and lowered into wells will keep long. He writes of the moon and vinegar, of cuttle ink, and of the glowworm. These are such facts as each one of us, divested of four centuries of learning, might try to record for posterity if we were the last of a dying race.

In a sense Francis Bacon was such a man. He was dying seemingly without scientific issue; the great continuity of learning for which he pleaded had been received indifferently by the world. Yet hidden in the *Sylva Sylvarum,* regarded as of little importance today, is a quite remarkable statement.

The passage is striking because it sets the stage for as pure a demonstration of the value of the induction for which Bacon argued as he could possibly have hoped for. Yet because two hundred and fifty years were to elapse in the reasoning process, men have forgotten the connection. After some observations upon changes in plants he remarks:

> The transmutation of species is, in the vulgar philosophy, pronounced impossible, and certainly it is a thing of difficulty, and requireth deep search into nature; but seeing there appear some manifest instances of it, the opinion of impossibility is to be rejected, and the means thereof to be found out.

This tolerant and studious observation with its evolutionary overtones was made before the nature of fossils was properly understood and before the length of geological time had been appreciated.

"The path of science," Bacon had proclaimed, "is not such that only one man can tread it at a time. Especially in the collecting of data the work can first be distributed and then combined. Men will begin to understand their own strength only when instead of many of them doing the same things, one shall take charge of one thing and one of another."

For the next two hundred years men allied in international societies originally foreseen by Bacon would make innumerable observations upon the strata of the earth, upon fossils, and upon animal and plant distributions. Heaps upon heaps of facts collected and combined by numerous workers would eventually lead to Darwin's great generalization. In the end Darwin himself was to write, "I worked upon the true principles of Baconian induction." The individual empirical observations which led to the theory of evolution and the recognition of human antiquity had been wrenched piecemeal from the earth.

Bacon, moreover, was keenly cognizant of the value of the history of science and philosophy. He deplored its neglect and urged that all such history of "oppositions, decays, depressions [and] removes . . . and other

events concerning learning will make men wise."
Even further, in *The Advancement of Learning* he indicates a complete awareness of what today we would
call the intellectual climate, or *Zeitgeist,* of an age.
He recognizes, and knows within himself, that it is
possible for able men to find themselves in an un-
congenial age. He advises men "to consider how the
constitution of their nature sorteth with the general
state of the times." If their natures and that of the
time are congenial they may allow themselves "more
scope and liberty." Otherwise their lives must "be
more close, retired, and reserved." There can be no
doubt that Bacon, for all his role in public office,
preferred consciously the latter course. His deep interest in history, his pioneer advocacy of studies in
comparative government, his plea for personal biogra-
phies of men other than princes, all reveal the breadth
of his horizon-circling mind.

Bacon also fully anticipated the folly of great
thinkers in their tendency to extrapolate too broadly
from the base of a single discovery. Thus, though he
admired Gilbert's discovery of the magnet, Bacon
wrote: "he has himself become a magnet: that is he
ascribed too many things to that force." One might
observe that this well-known tendency is apparent in
Louis Agassiz's final exaggeration of the extent of the

Ice Age, so that he envisioned it as covering the Amazon Basin. Other equally pertinent examples could be cited. Even the great thinker, or if not he, then his followers, sometimes show a tendency to create anew a ring within which they dance. Bacon's own emphasis upon induction was to suffer from similar misuse by his followers of the early nineteenth century who seized upon it as a device with which to castigate the evolutionists as "speculating" from insufficient accumulations of fact.

One other view which Bacon advocated was severely criticized by Macaulay and still evokes comment today—namely, his so-called neglect of genius. "The course I propose for the discovery of sciences," Bacon argues, "is such as leaves but little to the acuteness and strength of wits, but places all wits and understandings nearly on a level."

At first glance such a remark is apt to offend the superior intellect. In actuality, however, just as Clemenceau is reputed to have remarked that war was too important to be left to generals, so Bacon is not content to leave the development of the sciences to the sporadic appearances of genius. Nor, practically speaking, I think, would any educator today.

As noted in my first lecture, a careful reading of Bacon reveals that what he is anxious to achieve is the

49

triumph of the experimental method. This triumph demands the thorough institutionalization of science at many levels of activity. In one passage he encompasses in a brief fashion all those levels on which science operates today. "I take it," he writes prophetically, "that all those things are to be held possible and performable, which may be done by some persons, though not by everyone; and which may be done by many together, though not by one alone; and which may be done in the succession of ages, though not in one man's life; and lastly, which may be done by public designation and expense, though not by private means and endeavor." Only in this manner can the continuity of the scientific tradition be maintained and the small bricks which go into the building of great edifices be successfully gathered. "Even fourth rate men," Darwin was later to observe, "I hold to be of very high importance at least in the case of science."

If Bacon meant anything at all, he meant that working with the clay that sticks to common shoes was the only way to ensure the emergence of order and beauty from the misery of common life as his age knew it. He eliminated, in effect, reliance upon the rare elusive genius as a safe road into the future. It partook of too much risk and chance to rely upon such men alone. One must, instead, place one's hopes for Utopia

50

in the education of plain Tom Jones and Dick Thick-head.

Ironically, this was the message of a very great genius, an aristocrat who had lived all his life in the pomp of circumstance, but who, in the end, was willing to leave his name to later ages and his work to their just judgment. "I say without any imposture, that I . . . frail in health, involved in civil studies, coming to the obscurest of all subjects without guide or light, have done enough, if I have constructed the machine itself and the fabric, though I may not have employed or moved it."

Bacon had an enormous trust in the capacities of the human mind, even though no one had defined better than he its idols and distortions. "There be nothing in the globe of matter," he wrote, "that has not its parallel in the globe of crystal or the understanding." John Locke, almost a century later, is far more timid than this. Perhaps Bacon reposed too much hope in the common man. Or perhaps it is we who lack hope —the age for which Bacon waited being still far off, or a dream. But is it not a very great wonder that a man who spent all his life in the arrogant class-conscious court of a brutal age strove for personal power as a means of transmitting to the future an art which would, in a sense, make him, a very great genius, and men like him, less needed for human advancement?

I know of no similar event in all history. As an educator in a country which has placed its faith in the common man, I can only say that the serenity of Bacon's faith takes our breath away and gives him, at the same time, our hearts. For he, the Lord Chancellor, was willing to build his empire of hope from common clay—from men such as you and I. "It is not," he protests, "the pleasure of curiosity . . . nor the raising of the spirit, nor victory of wit, nor lucre of profession, nor ambition of honor or fame, nor inablement for business, that are the true ends of knowledge." Rather it is "a restitution and reinvesting of man to the sovereignty and power which he had in the first state of creation."

Those who dismiss Bacon as a scientist because he made no mechanical inventions have forgotten his own uncanny and preternatural answer, for he remarked that in the beginning there was only light. The grinding of machines and the sounds above us in the air we have taken for the scientific fruits he spoke of. After four hundred years the light, however, is only along the horizon—that beautiful, dry light of reason, which Bacon admired above all things, and which he spoke of as containing charity.

Ours cannot be the light he saw. Ours is still the vague and murky morning of humanity. He left his name, the name of all of us, to the charity of foreigners and the next ages. We presume if we think we are those addressed in his will. We are, instead, only a weary renewed version of the court he knew and the days he wore out in blackness. The gyrocompass in the warhead has no new motive behind it; the Elizabethan intrigues that flung up men of power and destroyed them have a too-familiar look; the religious massacres that shook Bacon's century have only a different name in ours.

There is something particularly touching about Bacon's growing concern "to make the mind of man, by help of art, a match for the nature of things." He knew, in this connection, that man the predator is also part of that nature man had to conquer in order to survive. Bacon had sat long in high places; he knew well men's lusts and rapacities. He knew them in the full violence of a barbaric age.

Although he has been accused of giving "good advice for Satan's kingdom," he understood from the beginning, and stated in no uncertain terms, that the technological arts "have an ambiguous or double use, and serve as well to promote as to prevent mischief and destruction, so that their virtue almost destroys or unwinds itself." "All natural bodies," he contended, with some dim, evolutionary foreboding, "have really two faces, a superior and inferior."

In one of those strange yet powerful sentences which project like reefs out of the sum total of his work, he gives us a Delphic prophecy: "Whatever vast and unusual swells may be raised in nature," he says, "as in the sea, the clouds, the earth or the like," so that in this age our mind flies immediately to man, "yet nature," he continues, "catches, entangles and holds all such outrages and insurrections in her inextricable net, wove as it were of adamant."

John Locke, some decades later, struck by the immensity of the great American forest, cried out, "In the beginning the whole world was like America." Had Bacon, spokesman for science, a hundred years earlier, seen the possibility of the return of that forest even before it had departed? Or did he look beyond this age to a time when, by greater art than now we practice, we may have made our peace with the nature of things?

"It must ever be kept in mind," Bacon urged, "that experiments of Light are more to be sought after than experiments of Fruit." The man was obsessed by light —that pure light of the first Morning of Creation before the making of things had commenced, before there was a garden and a serpent and a Fall, before there was strontium and the shadow of the mushroom cloud. He who will not attend to things like these can, in Bacon's own words, "neither win the kingdom of nature nor govern it."

Because Bacon saw and understood this light, it is well, I think, that he be not judged by us. Those who charge him, like a necromancer, with having called up from the deeps of time the direst features of the modern age, should ponder well his views upon the soul—"the world being in proportion," as he says, "inferior."

"By reason whereof," Bacon adds, "there is agreeable to the spirit of man, a more ample greatness, a more exact goodness, and a more absolute variety, than can be found in the nature of things."

The means to that goodness and those uses of life which Bacon sought for man can now be summarized. They are underlying precepts so firmly impressed upon our generation that we take them for granted, forgetting that, in the words of Kuno Fischer, "It was only Bacon's enthusiasm maintained through half a century, his dauntless tenacity - - - and his splendid powers of speech that gave to science wings."

First, Bacon wrote, "upon a given body to generate and superinduce a New Nature or Natures, is the work and aim of human power." He distinguished, in other words, between the lurking creativeness in nature and the natural world unsubjected to human influence. From "the beaten and ordinary paths of Nature" he would lead us to "another Nature which shall be convertible with the given Nature." Here, he indicates, lies a road to human power *"as in the present state of things human thought can scarcely comprehend or imagine."*[2] Over and over in one form or another he draws a distinction between *natura libera* and *natura vexata.* The last, of course, is nature vexed, put to the question, examined by means of experiment.

Second: As is evident from the human intrusion into nature exemplified above, Bacon, more than any other man of his epoch, recognized not alone the concealed novelty residing in nature, but also the unexploited power of humanity, not just to *live* in nature, but to create a new nature through the right use of human reason—"that which," as he expressed it, "without art would not be done." In this intensely imaginative act he recognized the creativity also implicit in the properly disciplined mind. He saw how the universe could be ingested by the human brain

[2] Italics mine. L. E.

and reflected back upon nature in new forms and guises. It is the great scope of this penetrating generalization, its recognition of the role of human culture both in nature and in history, which is so startling to discover at the dawn of the scientific era. This insight alone ranks Bacon far above the journeyman investigator, no matter how able or well-intentioned.

Last, and most importantly, perhaps, Bacon had a powerful sense of time. He looked upon it in a new way, even if he could not foresee its ultimate extent. "It may be objected to me with truth," he pleaded against his own mortality, "that my words require an age, a whole age to prove them, and many ages to perfect them." It has been said of primitive man, and could be said even of men not so primitive, such as our Elizabethan forebears, that they lacked the words and concepts to deal adequately with events remote in time or hidden behind the outer show of visible nature. It was Bacon's forceful defense of education—education *future*, not static or directed toward the past, that eventually swung people's heads about in a new direction. He was literally forcing man to grow conscious of his own culture by projecting an ideal yet dynamic version of it into the future. Inventiveness, creativity, lay in the future, not the past.

Bacon's human world can be seen as lying some-where between plain raw nature and Plato's world of disembodied ideal forms. It is the cultural world of man, subject, it is true, to confusion and contingency, but existing, so long as man exists, alongside of the physical world. This world is both disembodied and wonder-working. To produce its wonders it is only necessary for man in both reason and charity to turn his head toward the future rather than the past. Thus Bacon strives to make of man an actively anticipatory, rather than reminiscent or "present," creature. To anticipate, however, the human being must be made conscious of his own culture and the modes of its transmission. Education must assume a role un-guessed in his time and imperfectly realized in ours. It must neither denigrate nor worship the past: it must learn from it.

"Make," Bacon wrote, "the time to come the disciple of the time past and not the servant." The words ring with such axiomatic brevity that without reflection the sharp ax blade of his thought glances aside from our dull heads. Yet in that single phrase is contained the spirit of Bacon, the educator who, though trained in the profession of law, admonished posterity, "Trust not to your laws for correcting the times but give all strength to good education." No man in the long history of thought strove harder to lay his hands upon the future for the sake of unborn generations.

The words connote the essence of all that Bacon dreamed. Strangely, they are incorporated, no doubt imperfectly, in the life of a nation whose first colonizers left England in the *Sarah Constant* the year *The Advancement of Learning* was published. New India, he called us then. America was mountains and savages and untamed horizons. Bacon's ideas would lodge there like flying seeds and find suitable soil in the wilderness. A "new" and hopeful continent would arise from minds at work in the forest— a"Newfound Land," in Bacon's own words, "of inventions and sciences unknown."

BACON AND
THE MODERN DILEMMA

3.

Bacon and the Modern Dilemma

"I may truly say," wrote Bacon, in the time of his tragic fall, "my soul hath been a stranger in the course of my pilgrimage. I seem to have my conversation among the ancients more than among those with whom I live." I suppose, in essence, this is the story of every man who thinks, though there are centuries when such thought grows painfully intense, as in our own. Shakespeare—Bacon's great contemporary—again speaks of it from the shadows when he says:

"Sir, in my heart there was a kinde of fighting
That would not let me sleepe."

In one of those strange, elusive stories upon which the modern writer Walter de la Mare exerted all the powers of his marvelous poetic gift, a traveler musing over the quaint epitaphs in a country cemetery sud-

denly grows aware of the cold on a bleak hillside, of
the onset of a winter evening, of the miles he has yet
to travel, of the solitude he faces. He turns to go and
is suddenly confronted by a man who has appeared
from no place our traveler can discover, and who has
about him, though he is clothed in human garb and
form, an unearthly air of difference. The stranger,
who appears to be holding a forked twig like that
which diviners use, asks of our traveler, the road.
"Which," he queries, "is the way?"

The mundane, though sensitive, traveler indicates
the high road to town. The stranger, with a look of
revulsion upon his face, almost as though it flowed
from some secret information transmitted by the
forked twig he clutches, recoils in horror. The way—
the human way—that the traveler indicates to him, is
obviously not his way. The stranger has wandered,
perhaps like Bacon, out of some more celestial path-
way.

When our traveler turns from giving directions,
the stranger has gone, not necessarily supernaturally,
for de la Mare is careful to move within the realm of
the possible, but in a manner that leaves us suddenly
tormented with the notion that our road, the road to
town, the road of everyday life, has been rejected by
a person of divinatory powers who sees in it some

disaster not anticipated by ourselves. Suddenly in this magical and evocative winter landscape, the reader asks himself with an equal start of terror, "What *is* the way?" The road we have taken for granted is now filled with the shadowy menace and the anguished revulsion of that supernatural being who exists in all of us. A weird country tale—a ghost story if you will—has made us tremble before our human destiny.

Unlike the creatures who move within visible nature and are indeed shaped by that nature, man resembles the changeling of medieval fairy tales. He has suffered an exchange in the safe cradle of nature, for his earlier instinctive self. He is now susceptible, in the words of theologians, to unnatural desires. Equally, in the view of the evolutionist, he is subject to indefinite departure, but his destination is written in no decipherable tongue.

For in man, by contrast with the animal, two streams of evolution have met and merged: the biological and the cultural. The two streams are not always mutually compatible. Sometimes they break tumultuously against each other so that, to a degree not experienced by any other creature, man is dragged hither and thither, at one moment by the blind instincts of the forest, at the next by the strange intuitions of a higher self whose rationale he doubts and does not understand. He is capable of murder without conscience. He has denied himself thrice over, and is as familiar as Judas with the thirty pieces of silver.

He has come part way into an intangible realm determined by his own dreams. Even the dreams he doubts because they are not fanged and clawed like the life he sees about him. He is tormented, and torments. He loves—and sees his love cruelly rejected by his fellows. Far more than the double evolutionary creatures seen floundering on makeshift flippers from one medium to another, man is marred, transitory and imperfect.

Man's isolation is even more terrifying if he looks about at his fellow creatures and searches for signs of intelligence behind the universe. As Francis Bacon saw, "all things . . . are full of panic terrors; human things most of all; so infinitely tossed and troubled as they are with superstition (which is in truth nothing but a panic terror) especially in seasons of hardship, anxiety, and adversity."

Unaided, science has little power over human destiny save in a purely exterior and mechanical way. The beacon light of truth, as Hawthorne somewhere remarks, is often surrounded by the flapping wings of ungainly night birds drawn as unerringly as moths toward candlelight. Man's predicament is augmented by the fact that he is alone in the universe. He is locked in a single peculiar body; he can compare observations with no other form of life.

(He knows that every step he takes can lead him into some unexplored region from which he may never return. Each individual among us, haunted by memory, reveals this sense of fear. We cling to old photographs and letters because they comfort our intangible need for location in time. For this need of our nature science offers cold comfort. To recognize this, however, is not to belittle the role of science in our world. In his enthusiasm for a new magic, modern man has gone far in assigning to science—his own intellectual invention—a role of omnipotence not inherent in the invention itself. Bacon envisioned science as a powerful and enlightened servant—but never the master—of man.)

One of the things which must ever be remembered about Francis Bacon and the depth of his prophetic insight is that it remains, by the nature of his time, in a sense paradoxical. Bacon was one of the first time-conscious moderns. He felt on his brow as did no other man—even men more skilled in the devising of experiment—the wind of the oncoming future, those far-off airs blowing, as he put it in the language of the voyagers, "from the new continent." Ironically, as

67

we have seen, neither king, lawyer, nor scientist could tolerate Bacon's vision of the oncoming future. Because William Harvey was a scientist whose reputation has grown with the years, he is sometimes quoted by scholars even today as demonstrating that Bacon was a literary man who need not be taken seriously by historians of science.

That Bacon was a writer of great powers no one who has read his work would deny. He exercised, in fact, a profound stylistic influence both upon English writers who followed him and upon the scientists of the Royal Society. To say, for this reason, that he is of no scientific significance is to miss his importance as a statesman and philosopher of science as well as to deny to the scientist himself any greater role in discovery than the casual assemblage of facts. Harvey's attitude serves only to illustrate that great experimental scientists are not necessarily equally great philosophers, and that there may be realms denied to them. Similar able but particulate scientists, it could easily be pointed out, wrote disparagingly of Darwin in his time.

The great synthesizer who alters the outlook of a generation, who suddenly produces a kaleidescopic change in our vision of the world is apt to be the most envied, feared, and hated man among his contemporaries. Almost by instinct they feel in him the seed of a new order; they sense, even as they anathematize him, the passing away of the sane, substantial world they have long inhabited. Such a man is a kind of lens or gathering point through which past thought gathers, is reorganized, and radiates outward again into new forms.

"There are . . . minds," Emerson once remarked, "that deposit their dangerous unripe thoughts here and there to lie still for a time and be brooded in other minds, and the shell not to be broken until the next age, for them to begin, as new individuals, their career." Francis Bacon was such a man and it is perhaps for this very reason that there has been visited upon him, by both moralist and scientist alike, so much misplaced vituperation and rejection.

He has been criticized, almost in the same breath, as being falsely termed a scientist and, on the other hand, as being responsible for all the technological evils from which we suffer in the modern age. I have said that his vision was, to a degree, paradoxical. The reason lies in the fact that even the great visionary thinker never completely escapes his own age or the limitations it imposes upon him. Thus Bacon, the weather-tester who held up a finger to the winds of time, was trapped in an age still essentially almost static in its ideas of human duration and in the age and size of man's universe.

A man of the Renaissance, Bacon, for all his cynicism and knowledge of human frailty, still believed in man. He argues well and lucidly that to begin with doubt is, scientifically, to end in certainty, while to begin in certainty is to end in doubt. He failed to see that science, the doubter, might end in metaphysical doubt itself—doubt of the rationality of the universe, doubt as to the improvability of man. Today the "great machine" Bacon so well visualized, rolls on uncontrolled and infinitely devastating, shaking the lives of people in the remote jungles of the Congo as it torments equally the hearts of civilized men.

It is evident from his *New Atlantis* (1624), the Utopian fragment begun toward the end of his life and left unfinished, that this attempt to picture for humanity the state it might attain under science and just rule retains a certain static quality. Bacon is sure about the scientific achievements of his ideal state, but, after all, his pictured paradise is an island without population problems, though medicine is there, apparently, a high art. Moreover, like most of the Utopias of this period, it is hidden away from the corrupting influence of the world. It is an ideal and

70

moving presentation of men going about their affairs under noble and uplifting circumstances. It is, as someone has remarked, "ourselves made perfect."

But as to how this perfection is to come into being, Bacon is obscure. It is obvious that the wise men of the *New Atlantis* must keep their people from the debasing examples of human behavior in the world outside. Bacon, in other words, has found it easier to picture the growth of what he has termed "experiments of fruit" than to establish the reality of a breed of men worthy to enjoy them. Even the New Atlantis has had to remain armed and hidden, like Elizabethan England behind its sea-fogs.

The *New Atlantis* cannot be read for solutions to the endless permutations and combinations of cultural change, the opened doorway through which Bacon and his followers have thrust us, and through which there is no return. To Bacon all possible forms of knowledge of the world might be accumulated in a few scant generations. With education the clouded mirror of the mind might be cleansed. "It is true," he admitted in an earlier work, *Valerius Terminus*, "that there is a limitation rather potential than actual which is when the effect is possible, but the time or place yieldeth not matter or basis whereupon man should work."

71

In this statement we see the modern side of Bacon's mind estimating the play of chance and time. We see it again a few pages later when, in dealing with the logical aspects of contingency, he writes, "our purpose is not to stir up men's hopes but to guide their travels." "Liberty," he continues, speaking in a scientific sense, "is when the direction is not restrained to some definite means, but comprehendeth all the means and ways possible." For want of a variety of scientific choice, he is attempting to say, you may be prevented from achieving a scientific good, some desirable direction down which humanity might travel. The bewildering multiplicity of such roads, the recalcitrance of even educated choice, are not solely to be blamed on Bacon's four-hundred-year gap in experience. In fact, it could readily be contended that science, as he intended to practice it, has not been practiced at all.

Although he has been hailed with some justice as the prophet of industrial science, it is often forgotten that he wished from the beginning to press forward on all scientific fronts at once, instead of pursuing the piecemeal emergence of the various disciplines in the fashion in which investigation was actually carried out. Three centuries have been consumed in establishing certain anthropological facts that he asserted from

72

the beginning. As we have noted, he distinguished cultural and environmental influences completely from the racial factors with which they have been confused down to this day. He advocated the careful study and emulation of the heights of human achievement. Today scientific studies of "creativity" and the conditions governing the release of such energies in the human psyche are just beginning to be made. He believed and emphasized that it was within man's latent power to draw out of nature, as he puts it, "a second world."

It is here, however, that we come back upon that place of numerous crossroads where man has lifted the lantern of his intellect hopefully to many ambiguous if not treacherous sign posts. There is, we know now to our sorrow, more than one world to be drawn out of nature. When once drawn, like some irreplaceable card in a great game, that world leads on to others. Bacon's "second world" becomes a multiplying forest of worlds in which man's ability to choose is subdued to frightened day-to-day decisions.

One thing, however, becomes ever more apparent: the worlds drawn out of nature are human worlds, and their imperfections stem essentially from human inability to choose intelligently among those contingent and intertwined roads which Bacon hoped

73

would enhance our chances of making a proper and intelligent choice. Instead of regarding man as a corresponding problem, as Bacon's insight suggested, we chose, instead, to concentrate upon that natural world which he truthfully held to be Protean, malleable, and capable of human guidance. Although worlds can be drawn out of that maelstrom, they do not always serve the individual imprisoned within the substance of things.

D'Arcy Thompson, the late renowned British naturalist, saw, long after, in 1897, that with the coming of industrial man, contingency itself is subjected to a kind of increasing tempo of evolution. The simplicity of the rural village of Shakespeare's day, or even the complex, but stabilized and harmonious life of a very ancient civilization, are destroyed in the dissonance of excessive and rapid change. "Strike a new note," said Thompson, "import a foreign element to work and a new orbit, and the one accident gives birth to a myriad. Change, in short, breeds change, and chance—chance. We see indeed a sort of *evolution* of chance, an ever-increasing complexity of accident and possibilities. One wave started at the beginning of eternity breaks into component waves, and at once the theory of interference begins to operate." This evolution of chance is not contained within the human

74

domain. Arising within the human orbit it is reflected back into the natural world where man's industrial wastes and destructive experiments increasingly disrupt and unbalance the world of living nature.

Bacon, for reasons we need not pursue here, shared in some part with his age a belief in the biform nature of the worldly universe. "There is no nature," he says, "which can be regarded as simple; every one seeming to participate and be compounded of two. Man has something of the brute; the brute has something of the vegetable, the vegetable something of the inanimate body; and so, Bacon emphasizes "all things are in truth biformed and made up of a higher species and a lower." Strange though it may seem, in this respect Bacon, though existing on the brief Elizabethan stage of a short-term universe, was perhaps better prepared for the Protean writhings of external nature and the variability manifest in the interior world of thought than many a specialist in the material and biological sciences who would follow him.

Patrick Cruttwell, in his study of Shakespeare, comments on how frequently war within the individual, a sense of divided personality, is widespread in the spirit of that age, as it also is in ours:

"Within my soul there doth conduce a fight

Of this strange nature, that a thing inseparate,
Divides more widely than the skie and earth."

How much more we would see, I sometimes think,
if the world were lit solely by lightning flashes from
the Elizabethan stage. What miraculous insights and
perceptions might our senses be trained to receive
amidst the alternate crash of thunder and the hurtling
force that give a peculiar and momentary shine to an
old tree on a wet night. Our world might be transformed interiorly from its staid arrangement of laws
and uniformity of expression into one where the un-
expected and coruscating illumination constituted our
faith in reality.

Nor is such a world as incredible as it seems.
Physicists, it now appears, are convinced that a principle of uncertainty exists in the submicroscopic realm
of particles and that out of this queer domain of accident and impact has emerged, by some kind of mathe-
matical magic, the sustaining world of natural law by
which we make our way to the bank, the theater, to
our homes, and finally to our graves. Perhaps, after
all, a world so created has something still wild and
unpredictable lurking behind its more sober mani-
festations. It is my contention that this is true, and
that the rare freedom of the particle to do what most

76

particles never do is duplicated in the solitary universe of the human mind.

The lightning flashes, the smashed circuits through which, on occasion, leaps the light of universes beyond our ken, exist only in rare individuals. But the flashes from such minds can fascinate and light up through the arts of communication the intellects of those not necessarily endowed with genius. In a conformist age science must, for this reason, be wary of its own authority. The individual must be re-created in the light of a revivified humanism which sets the value of man, the unique, against that vast and ominous shadow of man the composite, the predictable, which is the delight of the machine. The polity we desire is that ever-creative polity which Robert Louis Stevenson had in mind when he spoke of each person as containing a group of incongruous and oft-times conflicting citizenry. Bacon himself was seeking the road by which the human mind might be opened to the full image of the world, not reduced to the little compass of a state machine.

It is through the individual brain alone that there passes the momentary illumination in which a whole human countryside may be transmuted in an instant. "A steep and unaccountable transition," Thoreau has described it, "from what is called a common sense view of things, to an infinitely expanded and liberating one, from seeing things as men describe them, to seeing them as men cannot describe them." Man's mind, like the expanding universe itself, is engaged in pouring over limitless horizons. At its heights of genius it betrays all the miraculous unexpectedness which we try vainly to eliminate from the universe. The great artist, whether he be musician, painter or poet, is known for this absolute unexpectedness. One does not see, one does not hear until he speaks to us out of that limitless creativity which is his gift.

The flash of lightning in a single brain also flickers along the horizon of our more ordinary heads. Without that single lightning stroke in a solitary mind, however, the rest of us would never have known the fairyland of the *Tempest,* the midnight world of Dostoevsky, or the blackbirds on the yellow harvest fields of Van Gogh. We would have seen blackbirds and endured the depravity of our own hearts, but it would not be the same landscape that the act of genius transformed. The world without Shakespeare's insights is a lesser world, our griefs shut more inarticulately in upon themselves. We grow mute at the thought—just as an element seems to disappear from sunlight without Van Gogh. Yet these creations we might call particle episodes in the human universe

—acts without precedent, a kind of disobedience of normality, unprophesiable by science, unduplicable by other individuals on demand. They are part of that unpredictable newness which keeps the universe from being fully explored by man.

Since this elusive "personality" of the particle may play a role in biological change and diversity, there is a way in which the mysterious world of particles may influence events within the realm of the living. It is just here, within the human domain of infinite variability and the individual act, that the role of the artist lies. Here the creative may be contrasted to the purely scientific approach to nature, although we must bear in mind that a man may be both a scientist and artist—an individual whose esthetic and humanistic interests are as much a part of his greatness in the eyes of the world as the technical skills which have brought him renown.

Ordinarily, however, there is between the two realms, a basic division which has been widened in the modern world. Granted that the great scientific discoverer may experience the esthetic joy of the true artist, a substantial difference still remains. For science seeks essentially to naturalize man in the structure of predictable law and conformity, whereas the artist is interested in man the individual.

"This is your star," says science. "Accept the world we describe to you." But the escaping human mind cries out, in the words of Chesterton, "We have come to the wrong star. . . . That is what makes life at once so splendid and so strange. The true happiness is that we don't fit. We come from somewhere else. We have lost our way."

A few years ago I chanced to write a book in which I had expressed some personal views and feelings upon birds, bones, spiders, and time, all subjects with which I had some degree of acquaintance. Scarcely had the work been published when I was sought out in my office by a serious young colleague. Now with utter and devastating confidence he had paid me a call in order to correct my deviations and to lead me back to the proper road of scholarship. He pointed out to me the time I had wasted—time which could have been more properly expended upon my own field of scientific investigation. The young man's view of science was a narrow one, but it illustrates a conviction all too common today: namely, that the authority of science is absolute.

To those who have substituted authoritarian science for authoritarian religion, individual thought is worthless unless it is the symbol for a reality which can be seen, tasted, felt, or thought about by everyone else.

80

Such men adhere as rigidly to a dogma as men of fanatical religiosity. They reject the world of the personal, the gay world of open, playful, or aspiring thought.

Here, indeed, we come upon a serious aspect of our discussion. For there is a widespread but totally erroneous impression that science is an unalterable and absolute system. It is supposed that other institutions change, but that science, after the discovery of the scientific method, remains adamant and inflexible in the purity of its basic outlook. This is an iron creed which is at least partly illusory. A very ill-defined thing known as the scientific method persists, but the motivations behind it have altered from century to century.

The science of the seventeenth century, as many historians have pointed out, was essentially theoretical and other-worldly. Its observations revolved largely about a world regarded as under divine control and balance. As we come into the nineteenth century, cosmic and organic evolution begin to effect a change in religious outlook. The rise of technology gave hope for a Baconian Utopia of the *New Atlantis* model. Problem solving became the rage of science. Today problem solving with mechanical models, even of living societies, continues to be popular. The emphasis, however, has shifted to power. From a theoretical desire to *understand* the universe, we have come to a point where it is felt we *must* understand it to survive. Governments expend billions upon particle research, cosmic-ray research, not because they have been imbued suddenly with a great hunger for truth, but for the very simple, if barbarous, reason that they know the power which lies in the particle. If the physicist learns the nature of the universe in his cyclotron well and good, but the search is for power.

One period, for reasons of its own, may be interested in stability, another in change. One may prefer morphology, another function. There are styles in science just as in other institutions. The Christianity of today is not totally the Christianity of five centuries ago; neither is science impervious to change. We have lived to see the technical progress that was hailed in one age as the saviour of man become the horror of the next. We have observed the same able and energetic minds that built lights, steamships, and telephones, turn with equal facility to the creation of what is euphemistically termed the "ultimate weapon."

It is in this reversal that the modern age comes off so badly. It does so because the forces which have been released have tended to produce an exaggerated conformity and, at the same time, an equally exaggerated assumption that science, a tool for manipulating the outside—our material universe—can be used to create happiness and ethical living. Science can be—and is—used by good men, but in its present sense it can scarcely be said to create them. Science, of course, in discovery represents the individual, but in the moment of triumph, science creates uniformity through which the mind of the individual once more flees away.

It is the part of the artist—the humanist—to defend that eternal flight just as it is the part of science to seek to impose laws, regularities and certainties. Man desires the certainties but he also transcends them. Thus, as in so many other aspects of life, man inhabits a realm half in and half out of nature, his mind reaching forever beyond the tool, the uniformity, the law, into some realm which is that of mind alone. The pen and the brush represent that eternal search, that conscious recognition of the individual as the unique creature beyond the statistic.

Modern science itself tacitly admits the individual, as in this statement from an English biologist: "We can be sure that, identical twins apart, each human being alive today differs genetically from any other human being; moreover, he is probably different from any other human being who has ever lived or is likely to live in thousands of years to come. The potential variation of human beings is enormously greater than their actual variation; to put it in another way, the ratio of possible men, to actual men is over-whelmingly large."

So far does modern science spell out for us that genetic indeterminancy which parallels, in a sense, the indeterminancy of the sub-atomic particle. Yet all the vast apparatus of modern scientific communication seems fanatically bent upon reducing that indeterminacy as quickly as possible into the mold of rigid order. Programs which do not satisfy in terms of millions vanish from the air. Gone from most of America is the kind of entertainment still to be found in certain of the world's pioneer backlands where a whole village may gather around a little company of visitors. The local musician hurries to the scene, an artist draws pictures to amuse the children, stories are told with gestures across the barrier of tongues, and an enormous release of creative talent goes on into the small hours of the night.

The technology which, in our culture, has released urban and even rural man from the quiet before his hearth log, has debauched his taste. Man no longer dreams over a book in which a soft voice, a constant companion, observes, exhorts, or sighs with him through the pangs of youth and age. Today he is more likely to sit before a screen and dream the mass dream which comes from outside. Strangely enough, our culture, in which man possesses far more leisure than that of any generation before him, is also so time-conscious that reading time on articles may be indicated in minutes, with the special "digest" magazines reducing the pabulum still further.

No one need object to the elucidation of scientific principles in clear, unornamental prose. What concerns us is the fact that there exists a new class of highly skilled barbarians—not representing the very great in science—who would confine men entirely to this diet. Once more there is revealed the curious and unappetizing puritanism which attaches itself all too readily to those who, without grace or humor, have found their salvation in "facts."

A hundred years ago there was violence in the world. The struggle for existence among living things was much written upon and it was popular for even such great scholars as Darwin and Wallace to dwell upon the fact that the vanquished died quickly and that the sum of good outweighed the pain. Along with the rising breed of scientific naturalists, however, there arose a different type of men. Stemming from the line of parson naturalists represented by Gilbert White, author of *The Natural History of Selbourne*, these literary explorers of nature have left a powerful influence upon English thought. The grim portrait of a starving lark cracking an empty snail shell before Richard Jefferies' window on a bleak winter day is from a different world entirely than that of the scientist. Jefferies' observation is sharp, his facts accurate, yet there is, in his description, a sense of his own poignant hunger—the hunger of a dying man—for the beauty of an earth insensible to human needs. Here again we are in the presence of an artist whose vision is unique.

Even though they were not discoverers in the objective sense, one feels, at times, that the great nature essayists had more individual perception than their scientific contemporaries. Theirs was a different contribution. They opened the minds of men by the sheer power of their own tremendous thought. The world of nature, once seen through the eye of genius, is never seen in quite the same manner afterward. A dimension has been added, something that lies beyond the careful analyses of professional biology. Something uncapturable by man passes over Hudson's vast landscapes. They may be touched with the silvery light from summer thistledown, or bleaker weathers, but always a strange nostalgia haunts his pages—the light of some lost star within his individual mind.

It is a different thing than that which some of my

colleagues desire, or that many in the scientific tradition appreciate, but without this rare and exquisite sensitivity to guide us, the truth is we are half blind. We will lack pity and tolerance not through intent, but from blindness. It is within the power of great art to shed on nature a light which can be had from no other source than the mind itself. It was from this doorway, perhaps, that de la Mare's celestial visitant had intruded. Nature, Emerson knew, is "the immense shadow of man." We have cast it in our image. To change nature, mystical though it sounds, we have to change ourselves. We have to draw out of nature that ideal second world which Bacon sought. The modern world is only slowly beginning to realize the profound implications of that idea.

Perhaps we can amplify to some degree, certain of our observations concerning man as he is related to the natural world. In Western Europe, for example, there used to be a strange old fear, a fear of mountains, precipices, of wild untrodden spaces which, to the superstitious heart, seemed to contain a hint of lurking violence or indifference to man. It is as though man has always felt in the presence of great stones and rarified air, something that dwarfed his confidence and set his thoughts to circling—an ice age, perhaps, still not outlived in the human mind.

There is a way through this barrier of the past that can be taken by science. It can analyze soil and stones. It can identify bones, listen to the radioactive tick of atoms in the substance of things. Science can spin the globe and follow the age-long marchings of man across the wastes of time and space.

Yet if we turn to the pages of the great nature essayists we may perceive once more the role which the gifted writer and thinker plays in the life of man. Science explores the natural world and thereby enhances our insight, but if we turn to the pages of *The Maine Woods,* regarded by critics as one of Thoreau's minor works, we come upon a mountain ascent quite unparalleled in the annals of literature.

The effect does not lie in the height of the mountain. It does not lie in the scientific or descriptive efforts made on the way up. Instead the cumulative effect is compounded of two things: a style so appropriate to the occasion that it evokes the shape of earth before man's hand had fallen upon it and, second, a terrible and original question posed on the mountain's summit. Somewhere along the road of that spiritual ascent—for it *was* a spiritual as well as a physical ascent—the pure observation gives way to awe, the obscure sense of the holy.

From the estimate of heights, of geological observation, Thoreau enters what he calls a "cloud factory" where mist was generated out of the pure air as fast as it flowed away. Stumbling onward over what he calls "the raw materials of a planet" he comments: "It was vast, titanic, and such as man never inhabits. Some part of the beholder, even some vital part, seems to escape through the loose grating of his ribs as he ascends. His reason is dispersed and shadowy, more thin and subtile, like the air. Vast, inhuman nature has got him at disadvantage, caught him alone, and pilfers him of some of his divine faculty." Thoreau felt himself in the presence of a force "not bound to be kind to man." "What is it," the naturalist whispers with awe, "to be admitted to a Museum, compared with being shown some star's surface, some hard matter in its home."

At this moment there enters into his apprehension a new view of substance, the heavy material body he had dragged up the mountain the while something insubstantial seemed to float out of his ribs. Pausing in astonishment, he remarks: "I stand in awe of my body, this matter to which I am bound has become so strange to me. I fear not spirits, ghosts, of which I am one—*that* my body might—but I fear bodies, I tremble to meet them. What is this Titan that has possession of me? Talk of mysteries!—think of our life in nature—daily to be shown matter, to come in contact with it—rocks, trees, wind on our cheeks! the solid earth, the actual world." Over and over he muses, his hands on the huge stones, "*Who* are we? Where are we?"

The essayist has been struck by an enormous paradox. In that cloud factory of the brain where ideas form as tenuously as mist streaming from mountain rocks, he has glimpsed the truth that mind is locked in matter like the spirit Ariel in a cloven pine. Like Ariel men struggle to escape the drag of the matter they inhabit, yet it is spirit that they fear. "A Titan grasps us," argues Thoreau, confronting the rocks of the great mountain, a mass solid enough not to be dragged about by the forces of life. "Think of our life in nature," he reiterates. "Who are we?"

From the streaming cloud-wrack of a mountain summit, the voice floats out to us before the fog closes in once more. In that arena of rock and wind we have moved for a moment in a titanic world and hurled at stone titanic questions. We have done so because a slight, gray-eyed man walked up a small mountain which, by some indefinable magic, he transformed into a platform for something, as he put it, "not kind to man."

I do not know in the whole of literature a more

90

penetrating expression of the spirit's horror of the substance it lies trapped within. It is the cry of an individual genius who has passed beyond science into a high domain of cloud. Let it not be forgotten, however, that Thoreau revered and loved true science, and that science and the human spirit together may find a way across that vast mountain whose shadow still looms menacingly above us.)

"If you would learn the secrets of nature," Thoreau insisted, "you must practice more humanity than others." It is the voice of a man who loved both knowledge and the humane tradition. His faith has been ill kept within our time.

Mystical truths, however, have a way of knowing neither time nor total death. Many years ago, as an impressionable youth, I found myself lost at evening in a rural and obscure corner of the United States. I was there because of certain curious and rare insects that the place afforded—beetles with armored excrescences, stick insects which changed their coloration like autumn grass. It was a country which, for equally odd and inbred reasons, was the domain of people of similar exuberance of character, as though nature, either physically or mentally, had prepared them for odd niches in a misfit world.

As I passed down a sandy backwoods track where I

hoped to obtain directions from a solitary house in the distance, I was overtaken by one of the frequent storms that blow up in that region. The sky turned dark and a splatter of rain struck the ruts of the road. Standing uncertainly at the roadside I heard a sudden rumble over a low plank bridge beyond me. A man high on a great load of hay was bearing down on me through the lowering dark. I could hear through the storm his harsh cries to the horses. I stepped forward to hail him and ask my directions. Perhaps he would give me a ride.

There happened then, in a single instant, one of those flame-lit revelations which destroy the natural world forever and replace it with some searing inner vision which accompanies us to the end of our lives. The horses, in the sound and fury of the elements, appeared, even with the loaded rick, to be approaching at a gallop. The dark figure of the farmer with the reins swayed high above them in some limbo of lightning and storm. At that moment I lifted my hand and stepped forward. The horses seemed to pause—even the rain.

Then, in a bolt of light that lit the man on the hayrick, the waste of sodden countryside, and what must have been my own horror-filled countenance, the rain plunged down once more. In that brief, momentary glimpse within the heart of the lightning, haloed, in fact, by its wet shine, I had seen a human face of so incredible a nature as still to amaze and mystify me as to its origin. It was—and this is no exaggeration—two faces welded vertically together along the midline, like the riveted iron toys of my childhood. One side was lumpish with swollen and malign excrescences; the other shone in the blue light, pale, ethereal, and remote—a face marked by suffering, yet serene and alien to that visage with which it shared this dreadful mortal frame.

As I instinctively shrank back, the great wagon leaped and rumbled on its way to vanish at what spot I knew not. As for me, I offer no explanations for my conduct. Perhaps my eyes deceived me in that flickering and grotesque darkness. Perhaps my mind had spent too long a day on the weird excesses of growth in horned beetles. Nevertheless I am sure that the figure on the hayrick had raised a shielding hand to his own face.

One does not, in youth, arrive at the total meaning of such incidents or the deep symbolism involved in them. Only if the event has been frightening enough, a revelation, so to speak, out of the heavens themselves, does it come to dominate the meaning of our lives. But that I saw the double face of mankind in that instant of vision I can no longer doubt. I saw man—all of us—galloping through a torrential landscape, diseased and fungoid, with that pale half-visage of nobility and despair dwarfed but serene upon a twofold countenance. I saw the great horses with their swaying load plunge down the storm-filled track. I saw and touched a hand to my own face.)

Recently it has been said by a great scientific historian that the day of the literary naturalist is done, that the precision of the laboratory is more and more encroaching upon that individual domain. I am convinced that this is a mistaken judgment. We forget—as Bacon did not forget—that there is a natural history of souls, nay, even of man himself, which can be learned only from the symbolism inherent in the world about him.

It is the natural history that led Hudson to glimpse eternity in some old men's faces at Land's End, that led Thoreau to see human civilizations as toadstools sprung up in the night by solitary roads, or which provoked Melville to experience in the sight of a sperm whale some colossal alien existence without which man himself would be incomplete.

"There is no Excellent Beauty that hath not some strangeness in the Proportion," wrote Bacon in his days of insight. Anyone who has picked up shells on

94

a strange beach can confirm his observation. But man, modern man, who has not contemplated his otherness, the multiplicity of other possible men who dwell or might have dwelt in him, has not realized the full terror and responsibility of existence.

It is through our minds alone that man passes like that swaying furious rider on the hayrick, farther and more desperately into the night. He is galloping—this twofold creature whom even Bacon glimpsed—across the storm-filled heath of time, from the dark world of the natural toward some dawn he seeks beyond the horizon.

Across that midnight landscape he rides with his toppling burden of despair and hope, bearing with him the beast's face and the dream, but unable to cast off either or to believe in either. For he is man, the changeling, in whom the sense of goodness has not perished, nor an eye for some supernatural guidepost in the night.

Bibliography

ANDERSON, FULTON H.
  Francis Bacon: His Career and Thought. University of Southern California Press, Los Angeles, 1962.
  The Philosophy of Francis Bacon. University of Chicago Press, 1948.
BAUMER, F. L.
  Religion and the Rise of Scepticism. Harcourt Brace, New York, 1960.
CROWTHER, J. G.
  Francis Bacon: The First Statesman of Science. Cresset Press, London, 1960.
DIXON, WILLIAM HEPWORTH
  Personal History of Lord Bacon. Ticknor and Fields, Boston, 1861.
EISELEY, LOREN C.
  "Francis Bacon as Educator," *Science,* Vol. 133, pp. 1197–1201. 1961.
FARRINGTON, BENJAMIN
  Francis Bacon: Philosopher of Industrial Science. Schuman, New York, 1949.
FISCHER, KUNO
  Bacon: His Philosophy and Time. London, c. 1857.
FRAZER, SIR JAMES
  The Golden Bough. 1 vol. ed. Macmillan, New York, 1942.

MERTON, ROBERT K.

"Singletons and Multiples in Scientific Discovery: A Chapter in the Sociology of Science," *Proceedings of the American Philosophical Society*, Vol. 105, pp. 470–486. Philadelphia, 1961.

NICHOL, JOHN

Francis Bacon: His Life and Philosophy. 2 vols. Blackwood, London, 1901.

PRIOR, MOODY E.

"Bacon's Man of Science," *Journal of the History of Ideas*, Vol. 15, pp. 348–370. 1959.

ROSTAND, JEAN

Error and Deception in Science. Hutchinson, London, 1960.

STORCK, JOHN

"Francis Bacon and Contemporary Philosophical Difficulties," *Journal of Philosophy*, Vol. 28, pp. 169–186. 1931.

THOMPSON, RUTH D'ARCY

D'Arcy Wentworth Thompson: The Scholar Naturalist 1860–1948. Oxford University Press, London, 1958.

WILLIAMS, CHARLES

Bacon. Harper & Brothers, New York, 1933.

Lightning Source UK Ltd.
Milton Keynes UK
UKHW020734140922
408851UK00005B/565

9 781376 165142